ABOUT THE *

Peter Killips was born in northwest London in 1932. Memories of early mornings recall awakening to the roaring of the lions in the London zoo, which was close by. Weekly visits to the zoo created a lifelong interest in wildlife, especially reptiles. A schoolboy friendship with an Indian, Anim, was the beginning of a fascination for India and its wildlife.

Killips has lived in Malaysia and Libya, and visited other places in the world always studying reptile behaviour in the field.

For many years he has been a member of the British Herpetological Society and the Leicestershire Natural History Society.

A Guide to the Flora and Fauna of Goa

P KILLIPS

Orient Longman

Dedicated to the memory of my son,
Mark, who died in a motor cycle
accident on 27 July 1979.

Orient Longman Limited
Registered office
3-6-272 Himayatnagar, Hyderabad 500 029 (A.P.), India

Other Offices
Bangalore/Bhopal/Bhubaneshwar/Calcutta
Chandigarh/Chennai/Ernakulam/Guwahati
Hyderabad/Lucknow/Mumbai
New Delhi/Patna

ISBN 81 250 1579 5

Maps by
Cartography Department
Sangam Books (India) Ltd., Hyderabad

Typeset by
Deepa Kamath
Kilpauk
Chennai 600 010

Printed in India at
NPT Offset Printers Pvt. Ltd.
Royapettah
Chennai 600 014

Published by
Orient Longman Limited
160 Anna Salai
Chennai 600 002

The coastline of India as depicted in the two maps in this book is
neither correct nor authentic.

FOREWORD

Pratapsingh Rane
Chief Minister of Goa

The beautiful state of Goa is the home for some of the most exquisite species of wildlife. This book is a vlauable contribution to wildlife literature, applicable only to the Goan species.

In this book the reader will find means of identifying the common and the unknown and thus it will create new interest in birds, animals and other wildlife and heighten the aesthetic senses of the readers by acquainting them with the astonishing range and splendour of our natural heritage.

I wish the author all the best in his future endeavours, and wish him all success in his current project with the Leicester University covering the study of reptiles and amphibians.

PRATAPSINGH RANE

PREFACE

The purpose of this book is to help the casual and interested visitor identify some of the common varieties of plants, birds and animals that they will see in Goa. All the photographs were taken on foot or from a car with a hand-held camera without the aid of tripod or hides. The results may lack the technical brilliance or the detail of more carefully contrived subjects but the intention was to illustrate each study as it is likely to be encountered by the casual observer.

Many opportunities were missed. Butterflies, common everywhere, are difficult to photograph. Mammals and reptiles, often seen crossing the road, are never willing to pose. Similarly, many birds were missed including the crested eagle which perched less than ten metres away, sat motionless for several minutes while I carefully changed my lens, and launched off into the valley the instant I attempted to focus. There was also the Indian robin and the bulbul which peered inquisitively at me through the window of the tourist office restaurant when I was without a camera. Such events were quite common.

Despite these limitations, I hope there is enough information provided here to help the casual visitor identify some of the natural sights in Goa.

ACKNOWLEDGMENTS

My sincere thanks to all who assisted:

Audrey Lomas and Nigel Killips of Leicestershire Museum

Mr Choudhury, (Former) Conservator of Forests, Goa

Mr Rohidas N. Naik, Deputy Director, Conservator of Forests, Wildlife Division

Mr Churchill Alamao

Mr John Fernandez

Lawrence, 'snakeman' of Panjim

The staff at the Mandovi Hotel and all friends, too numerous to mention, without whose help this book could not have been written.

The forest code on page 68 is drawn from the booklet issued by the Forest Department (Wildlife Division) Goa.

CONTENTS

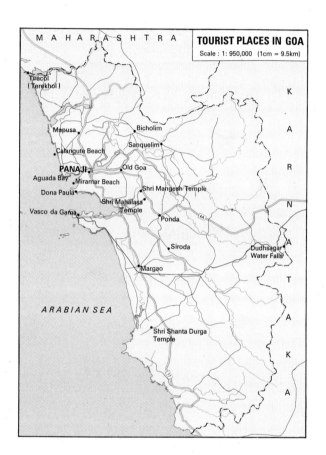

A note on
GOA

The world is divided into six zoogeographical zones, three of which are found in India. No other country or continent has more than two. India's forests and wildlife are rich and diverse. However, many habitats, and in consequence, their flora and fauna are now at risk due to human encroachment. But the situation is not without hope. The Indian government has established a Wildlife Division within its forest areas for the conservation of wildlife. Though this alone will not guarantee success it is a commendable step.

Goa was governed by the Portuguese for 450 years until they reluctantly relinquished control in 1961. It is positioned about midway on the west coast of India. It nestles between the Arabian Sea and the Western Ghats, a mountain range that runs north to south, approximately parallel to the coast from Kerala to Maharashtra. The rugged country and forests that grow on the Western Ghats are a haven for many species of birds and animals and perhaps, more vitally, provide a corridor along which they can migrate. The terrain also forms a natural barrier that deters human intrusion.

Most tourists come to sunbathe on the splendid beaches, and enjoy the delicious Goan food. Some visit buildings left by former colonists. Many are unaware that about ten per cent of the state's area is set aside as wildlife reserves and that a great variety of animals live in them.

The forests abound with various species of animals ranging from the huge gaur (900 kg) to the tiny field mouse (30 g). Bears and big cats hunt for their prey here. Birds are seen everywhere, usually unconcerned about human presence. Large butterflies attract attention by their size and colour. Lizards bask in the morning sun. Trees are seen everywhere: many are magnificent, some display brilliant flowers, others are of curious shapes and bear strange seed pods. In the market, different varieties of fruits and vegetables are aesthetically displayed.

Most tourists avoid the monsoon season although, as far as wildlife observation is concerned, this is actually the most interesting time. The vegetation takes on a fresh green hue and shallow pools appear after the first rains in otherwise arid areas. These pools seem to instantly come alive with dragonflies and water beetles. Later, birds feed on some of the hundreds of frogs that have come to spawn in their new homes. The rains are intermittent and are similar to the showers of a European spring but warmer.

There are several plant and animal species to be seen in Goa. Many of these will be encountered by the casual observer but a visit to a reserve will bring the visitor into contact with a wider variety of Goa's rich flora and fauna.

TREES, SHRUBS, FLOWERS AND FRUITS

There is an amazing variety of plant species growing in India. Many of them are native to India but several species were introduced years ago from other parts of the world. Most of these have become common now and have adapted themselves to Indian conditions. The species included here are mainly the more common varieties although a few more uncommon ones have also been described. They were selected because of their unusual shape or the brilliance of their flowers. Delicious fruits adorn some of the trees. Some of the trees grow to a great size and many of them have religious significance for the Hindus and the Buddhists. Parts of different plants are put to use in various ways. Several of the herbs are believed to have medicinal value.

1. COCONUT PALM
Cocos nucifera
This tree is a dominant feature of many tropical shores. Its fruit is used as a vegetable, and the water within is delicious. The flower can provide an alcoholic drink, toddy, and the leaves are used as thatch. The outer husk is a source of fibre from which ropes and similar produce are manufactured. In the picture, the *bayas* have suspended their nests from the leaves.

2. BOTTLE OR REGAL PALM
Roystonea regia
This tree is native to the West Indies and is commonly planted in parks and gardens. The smooth, whitish stems are stouter in the central portions giving the tree a bottle-like appearance.

4

3. FISH TAIL OR SAGO PALM

Caryota urens

The pith of this palm makes excellent sago hence its other name. The fibre obtained from this tree is called *kitul* locally and is used to make ropes, fishing lines and brushes. The leaves are used as elephant fodder.

4. TRAVELLER'S PALM

Ravenala madagascariensis

This tree is instantly recognisable by its banana like leaves radiating in an almost complete circle. The leaves may appear fringed but this is due to damage caused by wind.

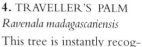

5. PALMYRA PALM
Borassus flabellifer
It is sometimes called the toddy palm as it is used to produce toddy, a cider like drink. Sugar is another product of this tree.

6. DATE PALM
Phoenix sylvestris
The fruits of this palm are smaller than the dates of West Asia. The fruit is very nutritious and has many medicinal applications.

6

7. BETEL NUT
Areca catechu
This tree was introduced from Malaysia. They are often planted in close proximity to each other to enable the collector of betel nuts to pass from one treetop to another.

8. BANYAN TREE
Ficus bengalensis
The banyan is sacred to Hindus and is easy to identify because of its breadth and the many aerial roots which hang down from the branches. These take root and enable the tree to spread out over a wide area.

9. INDIAN LABURNUM
Cassia fistula
This tree is particularly beautiful, when it flowers, with its cascades of fragrant yellow blooms. It is widespread in Goa.

10. MAST TREE
Polyalthia longifolia
This tree is so named as its trunk was once used to make masts. It is sacred to Hindus. The small, concealed purple fruits provide food for bats, monkeys and birds.

11. RAIN TREE
Samanea saman or *Pithecolobium saman*

Originally from South America it was introduced to India via Sri Lanka. It may grow over 15 m high and has a very broad spread, 20-25 m. The flowers are small and form fluffy white cones tinged with pink.

12. SILK COTTON TREE
Samalia malabarica

This tree flowers during January and February when it is usually leafless. Several species of birds and squirrels are attracted to the bright red flowers.

13. FLAME OF THE FOREST

Erythrina indica

These trees grow on the Taleigao Plateau, flowering on leafless branches, between January and March. The bark and seeds are believed to cure snakebite and scorpion stings.

14. GULMOHAR

Delonix regia

This beautiful tree is closely related to the Indian Laburnum.

15. SCARLET BELL TREE
Spathodea campanulata
This magnificent tree is native to Africa. It is not known when it was introduced into India. The inset shows a close-up of the flowers.

16. FRANGIPANI
Plumeria alba
This tree which is native to Mexico is now widespread in Goa. It blooms throughout the year and is often leafless. When picked, the strongly scented flowers are capable of surviving for a while without water. They are sometimes used in floral garlands and are generally planted near temples.

17. SCREW PINE *Pandanus tectorius*

This tree is often grown as a soil binder. The leaves are used to make mats and paper. The bright red leaf fibre is used for making fishing nets, fishing lines and sacking.

SHRUBS AND FLOWERS

18. BENGAL TRUMPET
— SKYFLOWER
Thunbergia grandiflora
This lovely lavender
hued flower blooms almost
throughout the year. It is
native to India.

19. WATER LILY *Nymphaea sp.*
This plant is commonly found growing in pools and lakes. The
Indian lotus (*Nelumbo nucifera*) looks similar but is easily distinguished
by its leaves which grow well above the water level, while those of
the water lily float on the surface.

20. WATER SNOWFLAKE *Nymphoides indicum*
This is a delicate plant that shares the habitat of the water lily but is otherwise unrelated.

21. LANTANA *Lantana aculeata*
This plant is native to Central America but has now overrun India. It is a straggling shrub with coarse, slightly prickly stems. Many butterflies find the clusters of orange flowers attractive. Birds eat the small dark berries, thus spreading the seed quickly.

22. PAPER FLOWER *Bougainvillea sp.*
This plant is native to South America. It is a semi climber often grown as a shrub. Bougainvilleas come in a magnificent range of colours, a far cry from the natural yellow or white variety found in Europe and the Americas.

23. CANDLE BUSH
Cassis alata
The bark of this plant is used in tanning, and the leaves and seeds have many medicinal uses.

24. DATURA
Datura stramonium

This is originally a Mexican plant now widespread throughout India. This plant is plentiful in Goa. It is believed to have many medicinal properties curing a wide range of ailments like ingrowing toenails, burns, sciatica and dandruff.

25. *Wigatea spicata*
This colourful climber can be seen growing on trees near the Mollem reserve.

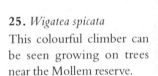

26. POWDER PUFF
Calliandra haematocephala
This beautiful shrub is grown in some hotels and public gardens. The flowers as you can see look like a 'powder puff', hence its name.

27. PEACOCK FLOWER *Caesalpinia pulcherrima*
This plant may have been native to South America. It grows very easily from seed to flower in one year.

28. MANGO
Mangifera indica
This tree is originally from the Burma–Assam region. It is now common throughout India. The tree in this picture is in flower and the inset shows an unripe fruit.

29. JACKFRUIT
Artocarpus heterophyllus
The fruit can weigh over 35 kg and is borne on the trunk. The wood of the tree is used in construction and also to produce a dye.

30. PINEAPPLE

Ananas comosus

This is a well known fruit-bearing plant. The leaf fibres are also useful. Because of the spiky nature of the plant it is often grown as a hedge.

31. PAPAYA (paw paw)

Carica papaya

This tree is native to Central America, and it is now widely grown in India. In addition to the orange, fleshy fruit which is delicious, the sap is said to contain a property which can tenderise meat. Male and female flowers are usually not found on the same tree.

32. CUSTARD APPLE *Anona squamosa*
This fruit is native to the West Indies. It is now cultivated all over India. It has a creamy texture and a delicious, vanilla custard flavour.

33. CASHEW *Anacardium occidentale*
This tree, related to the mango, was introduced by the Portuguese from South America. The harvested nuts are exported. The fruit is very tasty (but beware, the juice will permanently stain clothes), and is used to make a local drink—feni.

34. BREAD FRUIT
Artocarpus communis

This is a starchy fruit which is roasted and eaten. It is said that at the time of the 'Bounty' mutiny, Captain Bligh was commissioned by the British government to introduce the bread fruit plants to the West Indies from Tahiti.

35. BANANA
Musa paradisiaca

It is commonly grown in several parts of India for its delicious fruit. The other parts of the plant too have many uses. The flower (inflorescence) and pith are cooked and eaten. The trunk is used on auspicious occasions and the leaves are used to eat on (like a plate) and for packing food.

21

BUTTERFLIES

All the butterflies described in this section are common to Goa, though their presence will fluctuate according to season and even daily periods of activity. The specimens shown here are from a collection in the Leicester Museum (England). Most of the Indian varieties are represented here, but some sub species found in Goa, which have also been described, may vary slightly. Since the variations are very small the Goan representatives can be identified from the pictures used here. It should be noted that the museum specimens are mounted in a position that displays both fore and hindwings and a living insect may not necessarily adopt the same posture when it alights.

The butterflies selected for this book are of the larger varieties that will certainly be seen around Goa. There are many other smaller specimens—the blues, browns, yellows, whites and the skippers. When disturbed, the butterflies provide a brief kaleidoscopic display before re-alighting. Moths, some very large, can be observed at night.

36. SOUTHERN BIRDWING *Troides minos*
This is the largest butterfly in India. It is a common sight during the monsoon months (June to September) and for a month or two afterwards. It favours lantana bushes and is likely to be seen in the early mornings. It grows to a width of 140–290 mm.

37. CRIMSON ROSE *Pachliopta hector*
This common butterfly of southern India has bright red markings on its hindwings and body which are believed to act as a warning to birds. It grows to a width of 90–110 mm.

38. LIME BUTTERFLY *Papilio demoleus*
This butterfly is widely found across Central Asia, Southeast Asia and Australia. It grows to a width of 80–100 mm.

39. COMMON MORMON *Papilio polytes*
This is a large butterfly that closely resembles the rose flower. Note the enclosed white area on the hindwing. The female mimics the common crimson rose to protect itself from predators. It grows to a width of 90–100 mm.

40. COMMON ROSE *Pachliopta aristolochiae*
This butterfly is very similar to the common mormon. Note the white area expanding to the edge of the hindwing. It may be 80–110 mm in width.

41. COMMON BLUEBOTTLE *Graphium sarpedon*
This is a very distinct butterfly. When in flight only the bright iridescent green–blue is visible. It can be easily recognised and cannot be mistaken for any other species. It may be 80–90 mm in width.

42. TAILED JAY *Graphium agamemnon*
This butterfly is commonly seen in parks and gardens. The yellow
seen in the photograph is in fact, more of a fluorescent green.
It grows to a width of 85–100 mm.

43. COMMON WANDERER *Pareronia valeria*
This is the male of the species. The female is very similar to the
small blue tiger (see plate 49). Its width may be 65–80 mm.

44. COMMON JEZEBEL *Delias eucharis*
This is a common butterfly seen everywhere—even in cities.
It differs from most butterflies in having colours on its underwing.
Its width may be 65–85 mm.

45. Ventral side of the above.

46. PLAIN TIGER *Danaus chrysippus*
This butterfly belongs to the milkweed group—a family with worldwide representation. It feeds on poisonous plants and is therefore avoided by predators. It attains a width of 70–80 mm.

47. STRIPED TIGER *Danaus genutia*
This species, as its name suggests, bears prominent dark stripes. It closely resembles the common bluebottle butterfly (see plate 41). Its width varies from 75 to 95 mm.

48. COMMON CROW *Euploea core*
This butterfly can be identified by its gliding flight and its colour:
an iridescent blue over the brown on its wings which cannot be
seen in this photograph. Its width may be 85–95 mm.

49. BLUE TIGER *Tirumala limniace*
This butterfly is commonly seen in gardens. Its width may be
90–100 mm.

50. MALABAR TREE NYMPH *Idea malabarica*
This is one of the largest of Indian butterflies. It prefers a forest habitat and flies very slowly though it is capable of making a swift escape if necessary. It measures 110–160 mm in width.

51. PEACOCK PANSY *Junonia almana*
This is a common butterfly of the six member 'Pansy' group. It can be easily identified by its four eyespots, two small ones on its forewings and two large ones on its hindwings. Its width may be 45–60 mm.

AMPHIBIANS AND REPTILES

There are 4,553 known species of amphibians. Of these 4,000 are frogs and toads, 390 salamanders and 163 calcilians. Among the 140 genera of reptiles native to India, 67 are found in the Western Ghats. Indian reptiles range in size from the estuarine crocodile which grows to a length of 7 m to geckos less than 100 mm in length. There are approximately 45 different species of snakes in Goa and only five pose any danger to human life. These are the king cobra, a rare species which avoids human habitation, the common cobra, the common krait and the Russell's and saw scaled vipers. Almost all the serious incidents of snakebite in India may be attributed to the last four. Only a few specimens have been photographed as reptiles seldom allow themselves to be seen. You may catch a fleeting glimpse of a snake hurrying across the road. Geckos are often seen near a source of light feeding on insects attracted by the light. Geckos are harmless to humans. The same may be said of the tree frog and scurrying toads. The diurnal calotes run up the branchless trunks of palm trees stopping at intervals, head cocked seeking ants.

Crocodiles occasionally attack human beings and people do sometimes die from snakebite, but such incidents are fairly rare. They occur when attempts are made to kill or capture these reptiles or when they are trodden on. Seventy-two per cent of bites occur on the foot or lower leg, twenty-five per cent on the hand and arm. Most accidents occur in the evening or after dark. All reptiles avoid human contact and do not include them in their diet. They are not aggressive and if encountered by chance are no threat if left alone. In Goa, there are no records of crocodiles attacking humans.

52. COMMON TOAD
Bufo melanostictus
This toad can be found in any neighbourhood feeding on insects which gather under streetlights. It is less agile than a frog.

53. BALLOON FROG
Uperodon sp.
This frog has a short narrow head with creases of skin folded laterally. Its inflated body is used as a resonating chamber during the monsoon. This helps it to produce the distinct sound it makes in the rainy season.

54. INDIAN BULLFROG *Rana tigrina*
This is India's largest frog and is numbered among the large species of the world. It grows to a length of over 170 mm. This one measures over 200 mm weighing almost 700 g. Like many other amphibians, it is active during the monsoons and may be found in wells and tanks. If the water dries up the frogs bury themselves in the soil and await the next rains.

55. TREE FROG *Rhacophorus maculatus*
These frogs are sometimes found on walls of houses and even within rooms. The tips of their digits are formed into discs which enable them to climb walls. Their eggs are laid in foam nests which are formed when the frogs rub their hindlegs together during mating. When the tadpoles develop they 'drip' into the pool and are able to swim, thus avoiding the vulnerable larval stage of metamorphosis.

56. RUSSELL'S VIPER
Vipera russelli (venomous)
This is the 'speckled band' snake of the famous Sherlock Holmes story. This snake lives on rodents and is a valuable ally to the Indian farmer. It is of crepuscular habit and many of the incidents of snakebite are attributed to the Russell's viper. It grows to a length of 1.5–2 m.

57. INDIAN PYTHON
Python molurus
This snake is mainly nocturnal, but it remains alert during the day. Although terrestrial, it can climb trees. It thrives in a semi aquatic habitat. It feeds on mammals and birds (including poultry) and some species eat lizards. It grows to a length of 3–6 m.

58. COMMON COBRA
Naja naja (venomous)
This snake can be recognised by its erect, defensive position. In my experience this pose is not aggressive and if the snake is not pestered or further disturbed it will lower its head and disappear into the undergrowth. It grows to a length of 1–1.5 m.

59. VINE SNAKE
Ahaetulla nasutus
This snake is bright green, long and slender. It is difficult to spot in the shrubbery especially when it is still. It is the only known snake with a horizontal pupil. Its long snout has earned it the name 'eyepecker' in Sri Lanka and some parts of India, which is a misnomer. Harmless to humans it feeds on lizards. It grows to a length of 1.5–2 m.

60. SEA SNAKE *Lapemis curtus* (venomous)
These snakes are brought ashore in fishermen's nets. They are usually left to die on the beach. Although the sea snake is venomous its mouth is capable of only very slight distension making it an unlikely threat. It grows to a length of 85 cm.

61. RAT SNAKE *Ptyas mucosus*
This snake is usually two metres long, but sometimes grows to a length of three metres. It is a common diurnal snake and the casual observer is likely to encounter this species. It feeds on amphibians, lizards, small birds and mammals.

62. MARSH CROCODILE *Crocodylus palustris*
Of the three species of crocodiles in India the marsh crocodile is found in Goa. In addition to the marsh crocodile or 'mugger' there is the saltwater crocodile of the eastern estuaries and the thin-snouted gharial of the Ganges river system. All species are now uncommon. It grows to a length of 2.5–3.5 m.

63. BROOKS GECKO *Hemidactylus brooki*
This is the common house gecko. This lizard is often seen on the wall or ceiling, feeding on insects attracted to light. It grows to a length of 10–12 cm.

64. GARDEN LIZARD
Calotes versicolor
This lizard is not usually seen between November and February. The male has a red patch on the throat during the breeding season, and because of this, it is sometimes called the 'bloodsucker'. It grows to a length of 45 cm.

65. MONITOR LIZARD
Varanus bengalensis
This lizard can be seen basking on rocks or on open ground, and if alarmed will run at great speed. Its diet includes rodents, birds, amphibians, eggs and carrion. It grows to a length of 1.5 m.

66. STARRED TORTOISE *Geochelone elegans*
This is a very attractive tortoise. There is a black, radiating pattern on each yellow shield. It is crepuscular and is active during the monsoon season too. It is known to consume carrion. The female is about 38 cm in length and the male smaller, about 20 cm in length.

67. COMMON TERRAPIN *Melanochelys trijuga*
The tortoises belonging to this family differ from 'true' tortoises because they possess flattened rather than rounded limbs. They are nocturnal and on damp evenings may often be found grazing well away from water. As a defence mechanism, they produce an offensive odour when threatened.

BIRDS

Although the area of India is half that of the USA and about a third of Europe there are almost twice as many birds in India as in either of these countries. There are 1,250 bird species recorded in India many of which reside in Goa. Those illustrated in this section are birds that you will almost certainly see as you travel around Goa.

The egrets, herons, storks and kingfishers seek frogs and insects in the paddy fields. Occasionally, dozens of vultures circle in the sky. Above the towns and villages, kites constantly patrol, ever alert for a meal, always in competition with the crows.

Black drongos and tiny, green bee-eaters with pointed tails perch on telegraph wires and posts. The Indian roller sits sentinel-like, alert, awaiting unwary prey. Amid fluttering leaves, you may spy the spider hunters, ioras and leafbirds searching for hidden insects. There are many pleasant surprises around every corner and identifying these species will add pleasure and fulfillment to your stay in Goa.

68. CRESTED SERPENT EAGLE *Spilornis cheela*
This is commonly seen in the less populated areas and can be distinguished from the white-backed vulture by the dark band on the anterior portion of its underwing and the white band across its tail.

69. WHITE BACKED VULTURE *Gyps bengalensis*
Groups of these birds are a familiar sight and can be observed in many parts of Goa. It can be identified by its white back and rump.

41

70. ADJUTANT STORK *Leptopilus dubius*
This bird soars, vulture like, above the mangroves, feeds on fish and reptiles, and scavenges on carrion.

71. PARIAH KITE *Milvus migrans*
This aerial acrobat may be seen wheeling and diving often searching for food among refuse in towns and villages.

72. WHITE BREASTED KINGFISHER *Halcyon symrnensis*
A common sight, its reddish head contrasting with its prominent white breast distinguishes this from the similar-sized pied or black-capped kingfisher.

73. BRAHMINY KITE *Haliastur indus*
Extremely common, this kite is immediately recognised by its white head and reddish-brown body.

74. RED WATTLED LAPWING *Vanellus indicus*
These birds, often seen in pairs, are a common sight in marshes and on cultivated land. It has a distinctive 'noisy' call which is often repeated.

75. PURPLE MOORHEN *Porphyrio porphyrio*
Small groups of these birds can be found in marshes and reedbeds, a habitat they share with ducks, coots and grebes. Birdwatchers may visit Carambolim which is a well known habitat of waterbirds.

76. PADDY BIRD OR POND HERON *Ardeola grayii*
This bird can be seen in almost every marsh, pool or estuary. When in flight it is unmistakeable, its white wings contrasting with its brown body.

77. CATTLE EGRET *Bubulcus ibis*
This bird is commonly seen, particularly in association with a ploughing buffalo. Note its yellow beak and black feet.

78. REEF HERON
Egretta gularis
This bird is found near estuaries. Its plumage is sometimes white, but it can be distinguished from the cattle egret or little egret by its yellow beak and feet.

79. LITTLE EGRET
Egretta garzetta
This is a very common species which shares the habitats of the birds described earlier. Its black beak and yellow feet help to identify it.

80. INDIAN ROLLER *Coracias bengalensis*
This bird is similar in size and appearance to the European jay.
It is often seen sitting on posts looking for locusts, grasshoppers,
beetles and lizards on which it feeds.

81. JUNGLE MYNAH *Acridotheres fuscus*
The grey colour with the nasal crest distinguishes this mynah from
the related common mynah, which is more brown.

82. PEACOCK *Pavo cristatus*
Peafowl are found all over Goa except in residential areas. They are very shy, and are rarely seen. Their presence is revealed by their loud 'miaow' like call that can be heard over long distances. Their tails measure about 2 m in length.

83. BLUE TAILED BEE-EATER *Merops philippinus*
This bird, measuring 30 cm in height, is larger than the chestnut headed and small, green bee-eater. All species are commonly found in Goa.

84. GOLDEN ORIOLE *Oriolus oriolus*
The bright yellow plumage of this bird makes it quite easy to spot, particularly when in flight.

85. GREYHEADED MYNAH *Sturnus malabaricus*
This bird inhabits open country and sparse forests. It can be seen feeding on the nectar of flowering trees.

86. RED WHISKERED BULBUL *Pycnonotus jocosus*
This bird is quite common in woods, parks and gardens. It has a
beautiful song and is therefore often kept in homes.

87. CRESTED LARK *Galerida cristata*
This bird's crest and drab brown colour make it easy to identify.

88. RUFOUS BACKED SHRIKE *Lanius schach*

Shrikes habitually perch on prominent posts or branches while hunting insects and small reptiles. They can be easily recognised by the dark line running across their eyes and their heavy, hooked beaks.

89. BLACK DRONGO
Dicrurus adsimilis

These birds can be seen sitting, usually singly, on telegraph wires waiting for an insect to betray its presence.

MAMMALS

There are more than 500 different species of mammals found within the Indian region. Most mammals are nocturnal. It is therefore likely that they may only be spotted by accident. A mongoose or civet will sometimes cross the road ahead, or an otter may be spotted in an estuary or on a rocky shoreline. A birdlike chirrup close by may well turn out to be a small palm squirrel. Langurs or macaques stare back from their vantage points on a bush or building. Bats of many sizes fill the evening sky. At dusk, they can be seen emerging from beneath the eaves of a roof that has provided a daytime roost. By moonlight, flying foxes slowly flap their wings, silently searching for a meal of fruit.

To stand a chance of sighting bison, chital, sambar, wild boar or even a leopard or bear, a visit and an overnight stay at one of Goa's sanctuaries is essential. Wild elephants wander into the Mollem reserve from Karnataka and tigers have also been sighted here in recent years.

90. INDIAN FLYING FOX *Pteropus giganteus*

This large, fruit eating bat has a wing span of over 100 cm. It is usually seen flying before dusk with its distinctive slow wing beats. Daytime is spent at a roost. There is a large bat colony in Chorao.

91. CHITAL OR SPOTTED DEER *Axis axis*

This beautiful deer is mainly diurnal and lives in small herds.

92. SAMBAR *Cervus unicolor*
This is the largest Indian deer. A mature stag stands nearly 140–150 cm at the shoulder and can weigh 300 kg.

93. WILD BOAR *Sus scrofa*
The male boar stands 90 cm at the shoulder and its weight may be around 150–200 kg. This species is related to the European wild boar. Wild boars live in groups and feed on roots, crops, small reptiles and insects.

94. BONNET MACAQUE
Macaca radiata
Smaller than the langur, less than 60 cm in height when seated, this monkey has a pink face and an olive-brown body. Its underside is whitish-grey. The maximum weight this species can attain is 4 kg.

95. COMMON LANGUR *Presbytis entellus*
This monkey lives in small groups and can be distinguished from the macaque by its grey body, black face and slender build. It can attain a weight of 15 kg and a height of about 60–70 cm.

96. SLOTH BEAR
Melursus ursinus
This bear lives on fruits, roots and insects. More than 30 of this species were recorded at the last census (1993).

97. GAUR *Bos gaurus*
This is an impressive wild ox that can weigh 1000 kg and stands almost 2 m high at the shoulder. It can be found grazing at night, if you look for it with the aid of a powerful light.

98. INDIAN PORCUPINE *Hystrix indica*
This unmistakable mammal is nocturnal and may sometimes be
seen in the glare of car headlights.

99. SMALL INDIAN MONGOOSE *Herpestes auropunctatus*
This animal is usually spotted crossing the road. These occasions
rarely provide opportunities for photography. This particular
mongoose belongs to a friend who keeps it as a pet.

100. THREE STRIPED SQUIRREL

Funambulus palmarum

These squirrels are very common and inhabit gardens, coconut groves and forests. They make noisy chirruping calls. They grow to about 30 cm in length. In the Mollem reserve it may be possible to observe the metre long Indian giant squirrel (*Ratufa indica*).

MISCELLANEOUS

101. HOODED GRASSHOPPER *Teratodes monticollis*
This spectacular grasshopper is approximately 100 mm long and
15 mm high.

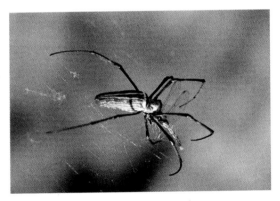

102. GIANT FOREST SPIDER *Nephila maculata*
This spider is commonly seen and can be observed even from a
vehicle passing through the forest. The female is 50 mm long and
15 mm wide. The diameter of the body including the legs is over
150 mm. The male is small and brown—a tenth the size of an adult
female. The web is large about 1.5–2 m in diameter. The spider in
this plate is eating a dragonfly.

103. SCORPION *Arachnid sp.*

This belongs to the same group in the animal kingdom as the spider and feeds on insects. The young ones are carried on the mother's back for about two weeks.

104. PRAYING MANTIS *Empusa sp.*

There are about two hundred species of praying mantis in India. They are green, brown or grey in colour and usually resemble the shrubs they frequent. The grey mantis shown here closely matches the colour of the wall.

105. INDIAN STICK INSECT *Carausius morosus*
This insect is mainly nocturnal and is so well camouflaged, that it is not easily seen. The females produce fertile eggs without the assistance of a male.

106. STARFISH *Asterias sp.*
This animal is a common sight on beaches. It is related to the sea urchin. Both these may be found at low tide.

107. MUD SKIPPER
Periphtalmus sp.

Mud skippers may be found among mangroves. They possess gills which are vestigeal and the insects breathe through their lungs.

108. FIDDLER OR CALLING CRAB *Uca sp.*

Huge colonies of these crabs live in burrows between high and low tide on muddy shorelines.

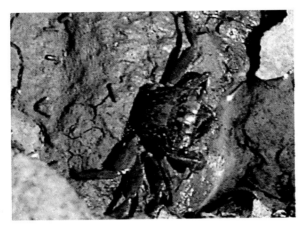

109. FIELD CRAB *Paratelphusa sp.*

This is a freshwater crab that may be found long distances from water—in forests, scrub and agricultural land.

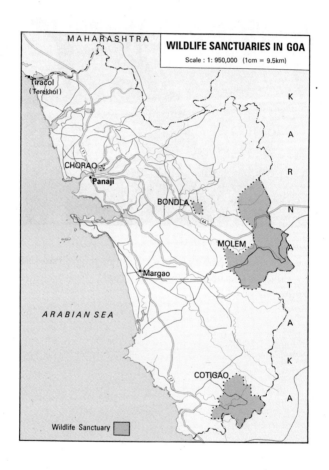

WILDLIFE SANCTUARIES IN GOA

Scale : 1: 950,000 (1cm = 9.5km)

MAHARASHTRA

Tiraçol
(Terekhol)

KARNATAKA

CHORAO
Panaji

BONDLA

MOLEM

Margao

ARABIAN SEA

COTIGAO

Wildlife Sanctuary

SANCTUARIES

There are three sanctuaries located along the foothills of the Western Ghats: the Bondla Wildlife Sanctuary, the Mollem National Park and the Cotigao Wildlife Sanctuary. At Bondla and Mollem accommodation is available set among stately trees and beautiful, delicate bamboos. Melodious birdsong can be heard echoing from within the forest, accompanied by the incessant chirruping of a million crickets. Shrill cicadas sometimes shatter the tranquillity of the evening. The animals found here include bonnet macaques and langurs, the massive gaur, the shy sambar and chital as well as the smaller barking deer and mouse deer. Black sloth bears, leopards and jungle cats, wild dogs and jackals, civets and mongoose can also be seen. Wild boar root for food on the forest floor, giant squirrels leap among the branches. Elephants and tigers are also frequently sighted.

There is yet another sanctuary, the Dr Salim Ali Bird Sanctuary. This differs from the others as it is set among mangroves on Chorao Island, close to Panaji, and is accessible only by boats provided by the Forest Department. There are otters, jackals and turtles, and there is also a large colony of flying fox and fruit bats in this sanctuary. Egrets and herons are also plentiful.

Early morning is the most rewarding time to tour the parks, although late evening visitors too will certainly find gaur, deer and wild boar. The Forest Department provides jeeps and guides. Watchtowers have been constructed at some sites. If you are using one of these it must be remembered that silence and immobility are vital to successful observation. Animals are unlikely to approach if they spot watchers. The slightest movement may give away your presence.

To check the availability of a jeep and to make arrangements for visits to any of the reserves make initial enquiries at the Office of the Conservator of Forests, Wildlife Division, Ground Floor, Junta House, Panaji. Tel: 229701

① SANCTUARY

Mollem National Park
and Bhagwan Mahavir Wildlife Sanctuary

Size: 240 sq km (approximately.)
Topography: hilly, dense forest in places
Location: located on National Highway 4A
  − 50 km east of Panaji
  − 55 km east of Margao
  − 6 km east of Mollem Railway Station
Transport: public transport buses to Bangalore, Hubli and Belgaum stop at the Mollem forest checkpost
Features of special interest: Dudsagar Waterfalls, Mahadeva Temple, Devil's Canyon and Sunset Point
Animals found: leopard, bear, wild boar, chital and gaur
Accommodation: cottages and dormitories
Booking: contact **The Tourist Information Service**, Mollem Tel: 600238

② SANCTUARY

Cotigao Wildlife Sanctuary

Size: 90 sq km (approximately)
Topography: some hills and plains with deciduous, evergreen and semi-evergreen forests on the eastern side
Location: located in south Goa 2 km off National Highway 17
  − 85 km south of Panaji
  − 50 km south of Margao
Transport: It is not easily accessible by public transport. Private transport will have to be arranged from Margao.
Features of special interest: a tree top house at Bhutpal, waterholes at Tulshimol and Dhantali

Animals found: leopard, bear, wild boar, sambar, chital and pangolin

Accommodation: Forest rest house, Polinguinim and Hotel Chaudi Canacona

Booking: contact **The Conservator of Forests**, 3rd Floor, Junta House, Panaji. Tel: 224747

❸ SANCTUARY

Bondla Wildlife Sanctuary

Size: 8 sq km (approximately)

Topography: hilly, forest area set on the foothills of the Western Ghats

Location: located on National Highway 4A
- 50 km east of Panaji,
- 40 km east of Margao

Transport: Public transport available upto Ponda and service buses or private taxies will have to be arranged upto the gate of the sanctuary.

Features of special interest: deer park and a wilderness trail with a watch tower and machan from where animals might be observed

Animals found: leopard, wild boar and sambar

Accommodation: cottages and dormitories

Booking: contact **The Manager**, Bondla Wildlife Sanctuary, Post Usgao, and **The Deputy Conservator of Forests**, Wildlife Division, Junta House, Panaji. Tel: 225926

❹ SANCTUARY

Dr Salim Ali Bird Sanctuary

Size: 2 sq km

Topography: mangrove forest

Location: located near National Highway 4A
  – 3 km east of Panaji
  – 35 km north of Margao

Features of special interest: set among the mangroves on Chorao Island near the mouth of the River Mandovi and an ideally-situated watchtower vantage point

Transport: No direct public transport available. Private transport and ferry services must be arranged from Old Goa.

Animals found: many estuarine birds, turtles, otters, jackal and flying fox

Accommodation: none at this site, but many suitable hotels in Panaji

Booking: open only on weekends; contact **Goa Tourism Development Corporation**, near Municipal Market, Panaji. Tel: 226515, 226728, 224132. Fax: 223926

FOREST CODE

- Sanctuaries are the homes of animals, the visitor is a guest. Behave as such.
- Animals have right of way.
- Keep to the track. Stay in the vehicle.
- Do not talk loudly, or play the radio or tape recorder. Never sound the horn.
- Do not throw away lighted cigarettes or matches.
- Do not leave embers or litter.
- Do not pick flowers or plants.
- Do not disturb animals, feed or chase them. Never approach the animals from the rear.
- Please report any contraventions or unusual happenings.
- Please assist in extinguishing any fire.
- Do not harrass or argue with the guide.

PLEASE ABIDE BY THESE RULES

CENSUS

A wildlife census was taken by the Forest Department (Wildlife Division) Goa, in 1984 and subsequently every three years. The census was discontinued after 1993.

SPECIES	YEAR			
	1984	1987	1990	1993
Indian bison	174	482	603	521
Sambar	45	71	162	315
Chital	87	50	156	421
Barking deer	40	97	106	104
Mouse deer	25	107	181	156
Wild boar	267	148	488	666
Giant squirrel	48	110	225	322
Sloth bear	3	18	20	34
Leopard	10	18	20	31
Tiger	—	—	—	3

FURTHER READING

Woodcock, Martin. 1980, *Collins Handguide to the Birds of the Indian Sub-Continent.* Collins.

Ali, Salim. 1983, *The Book of Indian Birds.* Bombay Natural History Society.

Prater, S.H. 1971, *The Book of Indian Animals.* Bombay Natural History Society.

Daniel, J.C. 1983, *The Book of Indian Reptiles.* Bombay Natural History Society.

Whitaker, R. 1978, *Common Indian Snakes: A Field Guide.* Macmillan India Limited.

Mukherjee, P. 1983, *Common Indian Trees.* (Nature Guides) Worldwide Fund for Nature, India.

Bole, P.V. Yogini, Vaghani. 1986, *Field Guide to the Common Trees of India.* Worldwide Fund for Nature, India.

Gay, Kehimkir, Punetha. 1992, *Common Butterflies of India.* (Nature Guides) Worldwide Fund for Nature, India.

Insight Guide to Indian Wildlife. APA Publications (Hong Kong) Ltd.